FOR BIRDS

Misha Maynerick Blaise is originally from Canada and was raised in the Colorado Rockies. She and her husband are the co-owners of a green building company and co-creators of two sons. They now call Austin, Texas, home. Misha is the author-illustrator of *This Phenomenal Life*, which has been translated into five languages and was a bestseller in China. She likes night swimming, music, researching stuff, and drinking black tea.

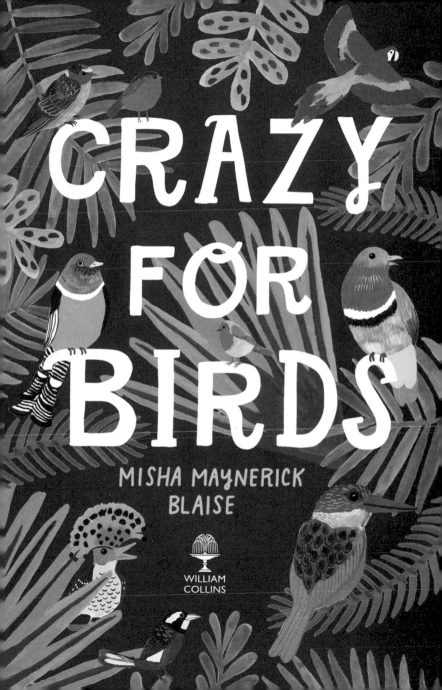

CRAZY FOR BIRDS

MISHA MAYNERICK BLAISE

WILLIAM COLLINS

William Collins
An imprint of HarperCollins*Publishers*
1 London Bridge Street
London SE1 9GF

WilliamCollinsBooks.com

First published in Great Britain by William Collins in 2020

10 9 8 7 6 5 4 3 2 1

First published in the United States by Penguin Books
An imprint of Penguin Random House LLC

A catalogue record for this book is available from the British Library

ISBN 978-0-00-839021-1

Printed and bound in Slovenia by the GPS Group

Typeset in Neutraface Text Demi
Designed by Sabrina Bowers

For Kazimir and Zarek

" ... In the GARDENS of UNITY
be birds of the SPIRIT,
SINGING of INNER TRUTHS
and MYSTERIES."

–THE BAHÁ'Í WRITINGS

YEMEN
LINNET

SECTIONS

HOODED
MERGANSER

Introduction

*Birds are the life of the skies,
and when they fly, they reveal
the thoughts of the skies.*

—D. H. Lawrence

THE WORLD ABOUNDS WITH A

Spectacular Diversity OF BIRDS:

feathered, egg-laying, extremely adaptable
winged vertebrates that live on all seven continents.
There are more than ten thousand living species
of birds in the world.

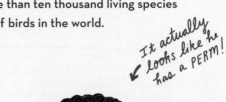

← It actually looks like he has a PERM!

THE (VERY GROOVY)
CURL-CRESTED
ARACARI

BIRDS LIVE VIRTUALLY **EVERYWHERE** ON **EARTH,** AND ENCOUNTERING THEM IS PART OF *The Universal human Experience.*

A few bird species can even be found on every continent (except Antarctica):

CAN FLY OVER 600 MILES A DAY.

BARN SWALLOW

A CLASSIC CUTIE.

MALLARD DUCK

ALSO FOUND in ANTARCTICA.

CAN LOCATE NEST BY SMELL IN THE DARK.

WILSON'S STORM-PETREL

THE MOST COMMON SPECIES of BIRD FOUND IN the WORLD is *The* CHICKEN.

There are about 20 billion chickens alive in the world at any time, which is about three chickens per human. The global chicken population outnumbers all of the planet's dogs, cats, rats, cows, and pigs, combined.

LIKE A BOSS.

One cool breed of chicken from Java is
AYAM CEMANI.

With the exception of its red blood, every other part of this chicken is black, including its tongue, internal organs, bones, skin, and eggs!

"THE GOtH Chicken"

On a daily basis, however, most urban dwellers
are far more likely to run into

A PIGEON.

It's estimated that there are about 400 million pigeons in the world, most of which live in cities.

Nikola Tesla was famously known to be enamored of the pigeons in his New York neighborhood. He carried a bag of feed with him and made pigeon friends everywhere he went. In his later years, having never married, he made his "feathery tribe" the center of his time and attention.

While he delighted in caring for all sick or injured pigeons, there was one female in particular whom he became obsessed with. When she was injured, he summoned his scientific powers to nurse her back to health. "Using all of my mechanical knowledge I invented a device by which I supported its body in comfort to let the bones heal."

"I LOVED that pigeon. I loved her as a man LOVES a WOMAN, and she LOVED ME. AS long AS I HAD her, there was MEANING to my LIFE."

-NIKOLA TESLA

Some write Tesla off as a crazy eccentric, but maybe in some primal way we all feel a kinship with birds.

HIGH FIVE!

FIST PUMP

HUMANS SHARE ABOUT 60 PERCENT OF OUR DNA WITH <u>ALL BIRDS</u>.

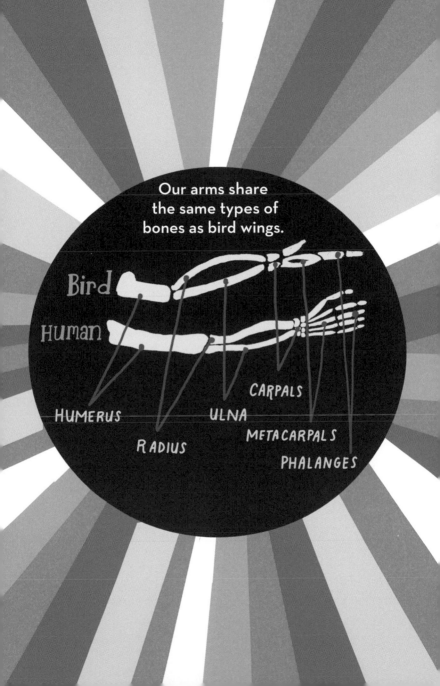

Even though mammals have evolved separately from birds for some 300 million years, scientists have found

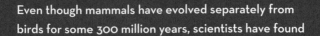

BIRDS and HUMANS
SHOW STRIKING PARALLELS
IN NEUROLOGICAL FUNCTIONING.

Avian brain mapping demonstrates that different parts of a bird's brain connect and interact with each other in ways similar to a human brain. For example, in both birds and humans, regions of the hippocampus (the area that is important for navigation and long-term memory) have dense connections to other parts of the brain, which implies that they function similarly.

Despite having small brains, birds have densely packed brain cells. In the parts of the brain where higher cognition takes place, some birds have as many neurons as smaller-brained primates.

A FEW EXAMPLES of BIRD INtELLIGENCE:

- The Eurasian Magpie passes the mirror test, a tool used to determine self-recognition.

- Corvids can craft and use tools, like bending a wire into a hook in order to grab something.

- Crows can distinguish between human faces, and hold a grudge against those who have harmed them in the past.

- Keas can use self-control and patience when working with a cooperative partner to solve intelligence tests.

But for most of us, it isn't scientific proof of intelligence that draws us to birds, but something far more primordial. Since ancient times, the flight of birds has lifted our eyes to the skies. Birds have connected us with the rhythms of the Earth: the passage of the day, the changing weather, the coming and going of seasons. They have inspired art and myth-making throughout the ages. Some speculate that human speech and music evolved

For millennia, humans evolved in close contact with birds and all of nature. The word

biophilia *

means a love of the living world; it describes the innate emotional bond that humans feel with other forms of nonhuman life. Nature is built into our cellular biology, and some part of us always longs to connect with it, the same way we long to spend time with a dearly loved friend.

* COINED BY BIOLOGIST E.O. WILSON

In today's high-tech, rapidly urbanizing world, many people feel deeply disconnected from the natural world—the air, the water, and the land—that sustains us. But the fact is we are always interacting with the wild and mysterious processes of the universe, no matter where we are.

Birds are ambassadors of this seemingly alternative realm of reality. The loud chirping of a bird in an urban car park reminds us that the natural world doesn't end when the city begins: The wilderness is not somewhere "out there"; it's right here, and we are totally connected to it.

Whether you take a cosmic mystical angle or a purely scientific approach, birds are an unending source of fascination. They connect us to who we truly are: members of an infinite, interconnected universe.

Birds are GOOD FOR OUR SOULS because they are part of the SOUL of the WORLD.

Eggs

I think that, if required on pain of death to name instantly the most perfect thing in the universe, I should risk my fate on a bird's egg.

—Thomas Wentworth Higginson, 1862, American Unitarian minister and abolitionist

BIRD *EGG* BASICS:

In each nesting attempt, the female of most avian species
lays one egg per day until she assembles
a group of eggs, called a clutch.

Female Turkeys
lay a CLUTCH of
up to 17 EGGS!

FYI:
A male Turkey
is called a
GOBBLER.

Each of these eggs is a miraculous self-contained life-support system that protects a single life within a hard shell; all eggs need in order to incubate is oxygen and warmth.

Interestingly, birds don't have to incubate their eggs right away. If they want, they can wait until a whole clutch is laid, and then incubate them all at one time so that their chicks will hatch together. The incubation period takes some birds a few days, others a few weeks.

BIRD EGGS ARE WONDERS of NATURE.

A stunning cosmic artistry is revealed through their diverse coloration and mathematically perfect form.

BROWN HAWK -OWL

This egg is an almost perfect sphere and looks like a ping-pong ball.

THE BLUE AND GREEN PIGMENT in bird eggs is also in HUMAN BLOOD and causes the green COLOR IN BRUISES.

GLOSSY IBIS

BANDED STILT

EUROPEAN GOLDEN PLOVER

NOW A POPULAR PAINT COLOR: "ROBIN'S EGG BLUE."

ROBIN

GUIRA CUCKOO

THIS IS THE SECOND LARGEST EGG IN THE WORLD. (OSTRICH EGGS ARE THE LARGEST.)

EMU

VICTORIA'S RIFLEBIRD

This egg's coating is GLOSSY and IRIDESCENT.

GREAT TINAMOU

BEE HUMMINGBIRD

THIS IS PERHAPS THE SMALLEST BIRD EGG IN the WORLD.

It's soooo CUTE!

It LOOKS AS IF It was painted by Jackson Pollock.

GREAT BOWERBIRD

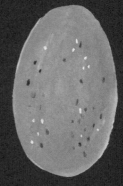

CASSOWARY

Birds have more color receptors in their eyes than humans (four photoreceptors, or cones, to our three), and therefore

they can see parts of the **ULtRAViOLEt LIGHt SPEctRUM** *that are totally invisible to us.*

WHOA.

Their cones also contain a droplet of oil that may allow them to see even more colors. So these already artful eggs may shine more brilliantly than we can possibly imagine.

Under ULtRAViOLEt light, chicken eggs glow a **VOLCANIC REd.**

THe gOLdCReST

is Europe's smallest bird. She cranks out a clutch of nine to twelve eggs, laying about one per day, which in total amounts to about one and a half times her own body weight. That's like a human giving birth to a new fifteen-pound baby every day for nearly two weeks.

NO BIG DEAL.

Some birds, like **THE BROWN-HEADED COWBIRD,** are "brood parasites." This winged brute will land in another bird's nest, knock out one of the eggs already laid there, and replace it with her own. The "foster" parent will in turn end up raising that Cowbird's chick, sometimes to the detriment of her own chicks. If that host bird rejects the new egg, the Cowbird will seek a violent revenge and destroy the host's whole clutch of eggs.

I HAVEN'T LOST MY TEMPER, IT'S RIGHT HERE!

BRUTAL!

One of the longest known incubation periods is that of

THE EMPEROR PENGUIN.

After laying the egg, the female carefully transfers the egg to the male (sometimes unsuccessfully), who will balance it on top of his feet for up to sixty-four days.

ANYONE GOT ANY NACHOS?

He does this together with a large group of other males, who huddle together to conserve heat during the long, dark winter. (Male Emperors have the longest known period of fasting of any bird during this time—some as long as 115 days!)

The Kiwi

has the largest egg to body size ratio of any bird—
its egg weighs up to 25 percent of its own body mass.

TIME FOR THE
XXL PREGNANCY
JEANS AGAIN.

The Ostrich

lays the world's largest egg (they weigh up to five pounds each, and have the same volume as about two dozen chicken eggs). But the egg only weighs 2 percent of the female's body weight, making it the world's smallest egg in proportion to its mother.

Both female and male Ostriches take turns incubating their eggs by sitting on them. Their eggs are laid in a dump nest, which is a communal nest where several Ostriches lay up to sixty eggs.

I DIDN'T EVEN REALIZE I WAS PREGNANT!

Ostriches can weigh up to four hundred pounds and grow as tall as nine feet. So basically an Ostrich egg needs to be strong enough to withstand the weight of a sumo wrestler sitting upon it.

BIRD EGGS ARE USED all over the world AS SYMBOLS of CULTURAL and RELIGIOUS SIGNIFICANCE:

Early Orthodox Christians in Mesopotamia dyed hard-boiled eggs red to symbolize the blood of Christ.

A hard-boiled egg is part of the Passover Seder plate in Judaism.

Ukrainian Pysanky is a batik-style painting tradition passed down from mother to daughter since the pre-Christian era.

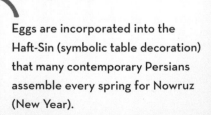

Eggs are incorporated into the Haft-Sin (symbolic table decoration) that many contemporary Persians assemble every spring for Nowruz (New Year).

Fabergé Eggs

are blinged-out jeweled eggs that the Russian czars commissioned jeweler Peter Carl Fabergé to create every Easter between 1885 and 1917. The fifty-seven surviving eggs each sell for millions.

REAL DIAMONDS!

Tsesarevich Egg 1912

AS A UNIVERSAL SYMBOL OF
RELIGIOUS RENEWAL,
OSTRICH EGGS HAVE BEEN USED TO DECORATE
ORTHODOX CHURCHES, SYNAGOGUES,
AND MOSQUES.

OSTRICH EGGS ON THE
TOP PILLARS OF THE
*Great Mosque of
Djenne,* MALI.

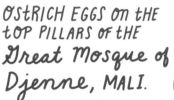

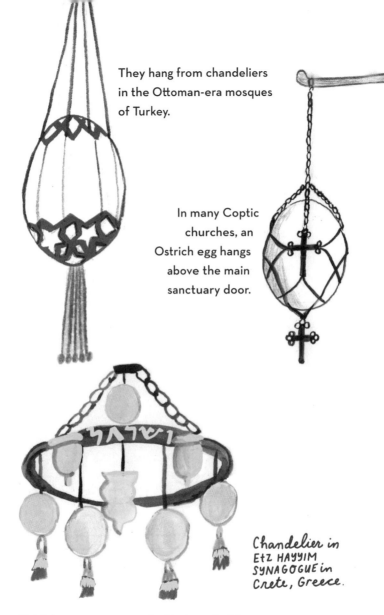

They hang from chandeliers in the Ottoman-era mosques of Turkey.

In many Coptic churches, an Ostrich egg hangs above the main sanctuary door.

Chandelier in Etz Hayyim Synagogue in Crete, Greece.

Ostrich eggs were reported to hang inside ancient Synagogues in Sicily and Yemen, and in Ottoman-era Turkey and Greece.

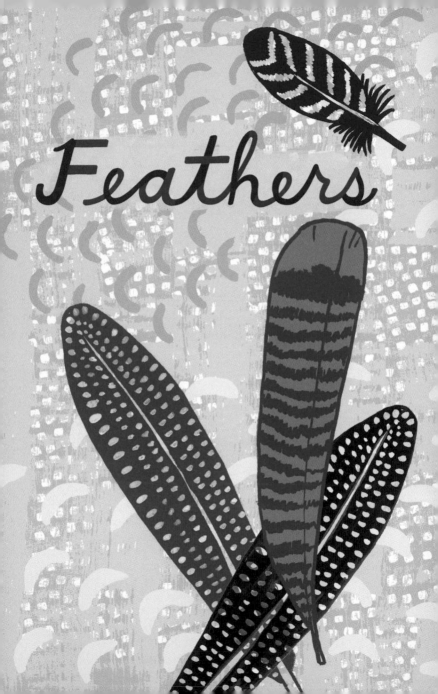

Feathers

Those who contemplate the beauty of the earth find reserves of strength that will endure as long as life lasts.

—Rachel Carson

BIRD FEATHER BASICS:

Feathers are some of the strongest, most lightweight, and flexible biological structures found in nature, and birds are the only living creatures on Earth that have them.

ALL FEATHERS HAVE the SAME BASIC STRUCTURE:

CENTRAL SHAFT
(RACHIS)

CALAMUS
the hollow shaft
at the base

Many famous documents, including the Magna Carta and the American Declaration of Independence, were written with quill pens made from flight feathers; the calamus is the part that holds the ink.

VANE

BARB

BARB

BARBULES

More than one thousand tiny barbules branch off of each barb, locking barbs together.

BARBICELS

Even tinier barbicels come off the end of each barbule to tightly seal barbs and barbules together. (This ingenious interlocking structure is what makes a feather water resistant.)

The average single flight feather is composed of more than one million tiny parts arranged in perfect alignment.

Rule of Thumb:

LADIES ARE DRAB;

MANDARIN
DUCK (Female)

GUYS ARE FAB!

JEALOUS?

MANDARIN DUCK (male)

Males tend to be more brightly colored than females, especially during breeding season.

There is no shortage of
eccentric plumage in the bird world:

ROYAL
FLYCATCHER

Ready to rock Rio's Carnival on
a moment's notice.

WINE-THROATED HUMMINGBIRD

All glammed up with his radiant
take on the hipster beard.

VICTORIA CROWNED PIGEON

His statement mohawk blends elegance
with a punk-rock sensibility.

EVEN the MOST Pretentious Fashionista CAN'T HELP BUT LOWER HER GUCCI GLASSES FOR A BETTER GLIMPSE OF THIS VULTERINE GUINEAFOWL'S PERFECTLY PATTERNED PLUMAGE.

The presence of the pigment melanin that determines the skin tone in humans also creates shades of brown, black, and even pale yellow in bird feathers. The more melanin that is in a feather, the darker it will appear, and the stronger and stiffer it will be. When the tips of flight feathers are black, the added strength helps reinforce the feather against wear and tear while flying.

SPOT-BELLIED
EAGLE-OWL

BROWN-HOODED KINGFISHER

But not all colors come from pigmentation. All blue and some green coloration comes from the way light particles interact with the actual structure of the feather. Iridescence appears when feather barbules refract light like a prism.

Spar-letta®
500 ml

STONEY®
TANGAWIZI

FEAth-ERS

are made of multiple layers of beta-keratin, the same protein that makes hoofs, horns, and our own human fingernails.

Feathers serve multiple purposes: They shield against UV light, insulate for warmth, provide camouflage, deflect water, and in most birds they are used for flight.

This CREEPY-CUTE TAWNY FROGMOUTH CHICK IS COVERED IN FUZZY, INSULATING DOWN FEATHERS.

Feathers are also used in sexual display, which John James Audubon could relate to. He was famously obsessed with his own "luxuriant" hair:

"MY locks FLEW FREELY from under my hat, and every LADY I met looked at them and then at ME until — SHE COULD SEE NO MORE."

John James Audubon was the artist who created *Birds of America*, one of the most expensive books in the world. In 2013 it was sold for $11.5 million (£9.3 million) at auction.

As if their scintillating array of tail feathers
wasn't enough to attract the ladies,

MALE

VIBRATE *their*
FEATHERS *to*
create AN
INFRASONIC
SOUND
that's INAUDIBLE
to humans,
BUT ALLURING
to PEAHENS.

A BALD EAGLE

relies on the precise coordination of its primary flight feathers to the extent that if it loses one, it will maintain balance by shedding a matching feather on the opposite wing.

This is deeply symbolic of something...

WHISKERED AUKLETS

use the elongated filoplume feathers around their eyes to navigate around volcanic rock passages in the darkness of night, much as cats use their whiskers as a sensory tool.

LIKE A CAT, BUT MORE FASHION FORWARD.

Flamingos

are grey when they are born, and turn pink only through a natural dye that is present in their diet of brine shrimp and blue-green algae.

A FLOCK of FLAMINGOS is Fabulously CALLED A Flamboyance.

MALE Club-Winged MANAKINS

of East Africa "sing" by rubbing their wings together, much like some insects (such as cicadas). By vibrating their wings together at a rate of more than one hundred beats per second (which is scientifically proven to be really freaking fast), they create a high-pitched "eeee!" sound.

eeeeeeee!

Does this relax you?

LIFE ISN'T PERFECT,
BUT YOUR FEATHERS CAN BE.

The male

Ribbon-tailed
Astrapia

has some of the longest tail feathers in the wild in
relation to its body size. They can grow up to one
meter (about three feet). In the bird world, females
may consider it a sign of fitness if a male can survive
while dragging around such ridiculously long plumes.

Probably the longest tail feathers in the world belong to

THE *Japanese* ONAGADORI,

a domesticated chicken breed whose tail feathers grow to astronomical lengths: They have been recorded at twelve meters (thirty-nine feet) long.

Millions of years of evolution have resulted in feathers being this totally excellent: Feathers were around even back during the time of the dinosaurs! Scientific consensus now points out that most dinosaurs (maybe all?) were feathered.

The Archaeopteryx is a descendent of the dinosaurs that is believed to be the world's earliest known bird.

Archaeopteryx

In fact, birds are considered to be living dinosaurs, relatives of Theropods that date back to the Mesozoic era. Around 66 million years ago, an asteroid caused the mass extinction of dinosaurs, but some 30 percent of living organisms survived, including a few birdlike dinosaurs—the ancestors of today's modern birds.

BIRDS LIKE THE SOUTHERN Cassowary
LOOK EERILY LIKE A FEATHERED DINOSAUR.

MONUMENTAL HEAD CREST

COLD, HARD REPTILIAN STARE

HAS BEEN KNOWN to KILL HUMANS.

FIVE-INCH CLAW ON FOOT

LOOKS EXACTLY LIKE A DINO CALLED *Corythoraptor Jacobsi.*

FEATHERS in HUMAN FORM

Indigenous and aboriginal people the world over have placed great significance on feathers, in the past and in the present:

TYN:

Elaborate feathered crowns worn by members of the Bamileke tribes of Cameroon; reserved for royal families, warriors, and kings.

ENKURARU:

An Ostrich feather headdress historically worn by a male Maasai warrior before completing the rite of passage of killing a lion.

DhARI / DhOERI:

The iconic headdress of the Aboriginal peoples of the Torres Strait Islands, Australia.

WAR BONNET:

Feathered headdress, usually made out of Eagle feathers, and worn by Native Americans belonging to some Great Plains tribes. They are traditionally reserved for males of a certain status and are considered very sacred.

FEATHERS ARE *decadently displayed* AROUND the WORLD *during* **MARDI GRAS** *and* **CARNIVAL** FESTIVALS.

Birds represent a link to both our
natural environment and to the possibility of
freedom to soar without boundaries.

—Rue Mapp, founder, Outdoor Afro

BIRD FLIGHT BASICS:

The wings of flight birds are perfectly designed to fly.
(Flight birds = birds that can fly)

HOOPOE

The bird's downstroke adds air pressure and creates thrust, which propels the bird upward. Air pushing against the curve of the wings creates lift. Bird bodies are so perfectly streamlined, it's easy for them to glide through the wind with little resistance.

CRESTED KINGFISHER

About half of the world's birds migrate, some even by foot or by swimming. Most birds migrate at night.

UNLIKE HUMAN MIGRANTS,
BIRDS AREN't HINDERED BY
IMMIGRATION CHECKPOINTS OR
BORDER WALLS.
THEY FLY FREELY ACROSS
NATIONAL BOUNDARIES, REFLECTING
THE *interconnected nature*
of **ALL ECOSYSTEMS** AND THE
ONENESS of *our* SHARED
PLANETARY HOMELAND.

Many species migrate on specific corridors called flyways that generally run between the north and south.

FOR EXAMPLE:
ABOUT FIVE BILLION (90 PERCENT) OF the BIRDS WHO BREED IN THE BOREAL FORESts OF NORTHERN CANADA MIGRATE SOUTH EVERY YEAR.

THE PALM WARBLER
Just one traveler on the north-south flyway.

Flyways FROM THE **BOREAL FORESTS:**

Western Flyway

Central Flyway

Eastern Flyway

THE MIGRATION JOURNEY IS BRUTAL

Birds are battered by storms when in flight, and they may find their stopover sites have been unexpectedly destroyed by humans. They may also discover themselves prey to domestic cats, one of the top human-caused threats to their existence. Migration is so dangerous that only half of the most migratory songbirds survive their route and are able to return the following breeding season.

IT IS NOT FULLY UNDERSTOOD HOW BIRDS NAVIGATE.

Research has shown that they use both natural and man-made landmarks and orient themselves at night with the help of the moon and the stars. Bits of iron in their inner ears may help them to orient themselves like a compass, and specialized proteins in the cells of their eyes allow them to actually see Earth's magnetic field.

THE ARCTIC TERN

is a freakishly fit, pocket-sized powerhouse that weighs about as much as a bar of soap (one hundred grams). Arctic Terns have one of the longest known migrations, flying an average of 70,900 kilometers (44,055 miles) from the Arctic regions to Antarctica (and then back again a few months later!).

EXAMPLE of FLIGHT PATH:

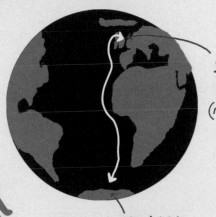

START:
THE FARNE
ISLANDS
(near England)

END: ANTARCTICA

CAN'T STOP, WON'T StOP.

BAR-TAILED GODWITS

gorge themselves for weeks until they double their body weight. The excess chub enables them to survive as they fly some 11,000 kilometers (7,000 miles) NONSTOP, without eating or drinking, from the Arctic (Siberia or west Alaska) to eastern Australia or New Zealand. The ultra-intense journey takes about eight days.

Sometimes a bird has to make a few adjustments before it can take flight.

you MIGHT NOT WANT TO SEE THIS.

When an

ANDEAN CONDOR

happens upon carrion, it will binge eat several pounds of meat. If it gets too heavy to fly away in an emergency, it will puke up whatever is needed to lighten its load.

COMMON SWIFTS ARE TRUE EXTREMISTS;

they spend **TEN MONTHS** OF the NONBREEDING SEASON **IN THE AIR!**

ARE WE THERE YET, MOM?

ONLY EIGHT MORE MONTHS, HONEY.

These birds eat, drink, and do everything else on the wing. Even the lazy ones who stop occasionally to rest for a few minutes still spend 99 percent of their time in flight.

How can swifts stay aloft for so long? Research has shown that birds can rest one hemisphere of their brain at a time while flying. Unbelievably, they can even shut down both sides and enter REM, a state characterized by total muscle relaxation, while coasting on the wind.

THE BAR-HEADED GOOSE

IS AMONG THE HIGHEST-FLYING BIRDS,

passing directly above the highest peaks of the Himalayas, upward of 8 kilometers (27,000 feet) on its migration route. (That's higher than a helicopter can fly!) The birds have to withstand frigid temperatures as low as –50 C (–58 F) and extremely low air pressure at that altitude.

One of the most mysterious ways that birds inhabit the skies is called a

MURMURATION.

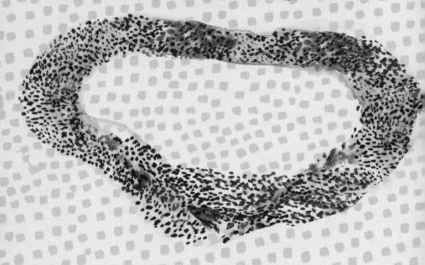

Thousands of individual starlings fly within inches of one another, creating a shimmering mass of birds that expands and contracts, forming ever-moving free-form shapes across the sky.

THE PEREGRINE FALCON

IS A RECORD HOLDER FOR *Speed.*

It flies to extreme heights and then drops down (a motion called stooping), tucking its wings in to become an aerodynamic bird bullet. One was recorded speeding through the air at more than 380 kilometers per hour (236 miles per hour).

THE POWER of FLIGHt

The ARt of FALCONRY:

For thousands of years, humans around the world have harnessed the power of bird flight by using birds of prey, such as Falcons, Hawks, and Eagles, as hunting partners.

UNESCO includes falconry on its list of Intangible Cultural Heritage of Humanity. Eighteen countries are recognized in the falconry category, making it one of the most recognized living cultural practices that unite humanity.

Homing PIGEONS

have an incredible sense of
direction and have been trained for
millennia to deliver messages.

(HAT OPTIONAL)

MAIL

← MESSAGE

The ancient Greeks used pigeons to spread news of the Olympics.
In the thirteenth century, Genghis Khan used a pigeon post system
throughout Asia and Europe.

During World War I, carrier pigeons were delivered to the front lines in baskets. Once there, soldiers would attach messages to the birds, who would then fly "home" to their coops behind the lines.

During World War II, a military unit called the United States Army Pigeon Service consisted of about 54,000 war pigeons. The birds had a 90 percent success rate in transporting messages.

THE DICKIN MEDAL

The People's Dispensary for Sick Animals in the United Kingdom awarded thirty-two pigeons this medal for animal bravery during World War II. War heroes included Winkie, Tyke, Mercury, Gustav, Mary, and Paddy.

BIRD FLIGHT SYMBOLIZES *Freedom* AND THE *Ascension of the soul.*

For centuries Zoroastrians carried out a funeral rite
in which they lay their dead outside in
a raised structure called

A TOWER of SILENCE.

Bodies were left there to be consumed by birds of prey,
particularly vultures. This offering of one's body as
nourishment for the birds is seen by Zoroastrians as a final
act of charity. (This tradition is still practised today among
some groups, such as the Parsis of India.)

*The natural world is the
greatest source of excitement.
The greatest source of visual beauty.
It is the greatest source of so much in life
that makes life worth living.*

—Sir David Attenborough

BIRD LOVE BASICS:

Birds display some of the most
bizarre and eccentric mating
rituals in the natural world.

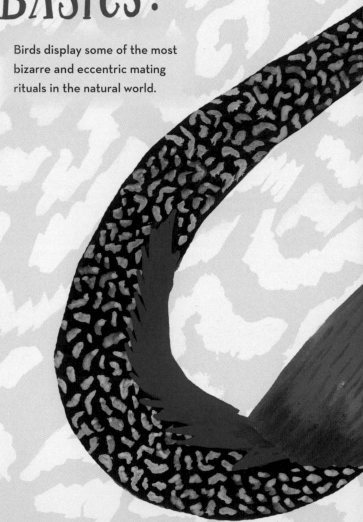

THE
GØLDEN
PHEASANT:
A man of
MYSTERY

And after all of the (usually) male's exhaustive work, most bird couples only need a few seconds to achieve successful fertilization. But those few seconds hold the key to species survival, something birds take very seriously.

THIS MALE FRIGAtEBIRD INFLAtES A HUGE RED THROAT POUCH AS PARt OF HIS COURtSHIP DISPLAY.

Every breeding season, male birds experience a rush of testosterone that triggers them into a state of eccentric spectacle. They ramp up on color, feather length, and overall swagger in an attempt to impress the ladies.

Like the
horn on a unicorn,
but more <u>MAGICAL</u>.

The male

TEMMINCK'S TRAGOPAN,

already a handsome little pheasant, unexpectedly unfurls a bib of breathtaking color and pattern. As two rubbery horns sprout up from his head, he metamorphoses into an almost otherworldly creature.

This is not a CHUBBY deer from another dimension; this is a REAL-LIFE bird.

THE *Superb* Bird-of-Paradise

takes an even more psychedelic approach. Starting out as a pretty basic-looking black bird, he stretches his wings out and curves his head back to transform into what appears to our eye as a glowing smiley face.

I'm weirdly feeling very attracted to this.

TRUE STORY:
THE GREATER SAGE-GROUSE

PUFFS OUT TWO GIANT AIR SACS ON ITS CHEST, THEN JIGGLES THEM UP AND DOWN UNTIL THEY SMACK TOGETHER. THIS IN TURN CREATES A LOUD **"BLUP!"** SOUND THAT CAN BE HEARD UP TO THREE KILOMETERS (1.9 MILES) AWAY.

IF YOU'VE GOT IT, FLAUNT IT.

The male
LONG-TAILED MANAKIN

has a wingman who assists with his courtship dance.

At one point, the two males jump rapidly in a hypnotic circular motion, one replacing the other as if on a carousel, singing out a bizarre, buzzy call on each jump. In the end, the alpha always gets the lady, and his devoted helper waits until his services are needed again.

SOME MALE
BOWERBIRDS
SEEK to IMPRESS FEMALES with their SOPHISTICATED DESIGN-BUILDING SKILLS.

The male meticulously sculpts a structure of sticks, called a bower, and decorates it with brightly colored elements, both natural and man-made. It can't be overstated what a perfectionist this bird is. He is singularly obsessed with his beloved bower.

I MADE THIS FOR ME,
I MEAN YOU.

Ornithologists have even found that if they move items around in his bower, he will quickly notice the upset and return everything to normal. Even if he succeeds in mating with a female, an instant later the affair is forgotten and he is back to fussing with his prized architectural masterpiece.

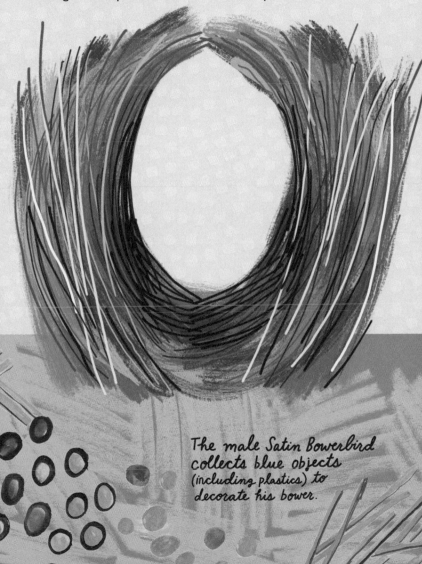

The male Satin Bowerbird collects blue objects (including plastics) to decorate his bower.

THE MALE *Flame* *Bowerbird*

mesmerizes his female visitor by alternately expanding and contracting his pupils. He then makes a raspy, asthmatic call, stares intently at his extended wing, and executes a series of sensuous, slow-motion shoulder rolls.

KKHHHey lady.

THE MALE BLUE-FOOTED Booby

does an awkward, exaggerated march in which he flaunts his feet to the female.

PREPARE to be AMAZED.

So DAZZLING!

Keeping in mind that birds see a much wider spectrum of color than we do, who knows what kind of enchanting display of blue magic the female sees emanating from those feet.

SOME MALE **RUFFS** ARE CROSS-dRESSERS;

they have a genetic trait that allows them to imitate the female's appearance and song. Males gather together and form what's called a lek, a sort of round-up that some bird species use so that males can strut their stuff in front of a female jury.

While the majority of the males in the lek are eagerly trying to outdo one another in a show of puffed-up braggadocio, the cross-dressed Ruff strategically saunters by undetected, then quickly makes moves on the lady of his choice.

YOU CAN TRUST ME.

THE MALE RED-BACKED FAIRYWREN

has a puberty-like transition around age two.

BEFORE

(LACKLUSTER)

AFTER

(STONE-COLD HUSTLER)

THE MALE KING-of-SAXONY BIRD-OF-PARADISE

has two stiff feathers that emerge from his head like extremely long antennae, coupled with a call that sounds like sci-fi laser beams.

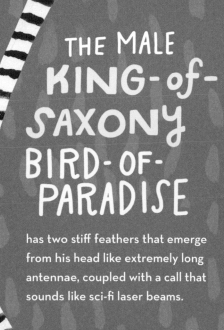

pew pew!

Cute couples:

Some 90 percent of the world's birds form monogamous pair bonds that can last one nesting, one breeding season, or even for life. How special for these little lovebirds!

I'm YOURS FOREVER

...or at least right now.

But THE REAL DRAMA StartS WHEN THE DNA testing BEGINS.

In most species, eggs in the nest of a supposedly monogamous pair can be traced to one or more other fathers!

Songbirds are known to be the most BRAZEN PHILANDERERS, with a whopping 86 percent of their species CHEATING on THEIR PARTNERS.

A FEMALE Jacana

assembles a harem of up to five males with whom she mates. As soon as she lays eggs with one guy, she is off for the next, and the male is left behind to incubate the eggs and raise the young.

I'LL JUST WAIT HERE FOR YOU.

SEE YOU LATER?

SO.... WHERE ARE YOU GOING?

JACANAS HAVE EXTREMELY LONG TOES (FOR BALANCING ON LILYPADS).

The national bird of Peru,

THE ANDEAN
COCK-OF-THE-ROCK,

spends his time perfecting his fresh dance moves and haunting
mating calls (he would blend in nicely with a horror film soundtrack).
After mating, he skips out on any boring domestic duties and heads
back to the lek for more action, confident he is the rockingest cock
in all of South America.

ICONS of LOVE

Some birds make the perfect
symbol for human relationships.

The dove is associated with
APHRODITE,
*THE GREEK GODDESS
of LOVE,* and has since
become a staple symbol for
Valentine's Day.

Parrots were considered symbols of LOVE in ancient India.

In the Kama Sutra, one of the sixty-four requirements of a man was to teach a Parrot or a Mynah how to talk.

The Chinese proverb

"Two Mandarin Ducks playing in water"

refers to a couple in love.

Mandarin Ducks are also symbols of love and marriage in Japan and South Korea.

Wherever there are birds,
there is hope.

—Mehmet Murat Ildan

BIRD NEST BASICS:

For many bird species, nests are really just temporary structures to hold eggs. They don't live in their nests, and when breeding season is over, they are on the move again.

What's amazing about bird nests is that birds make them using their **BEAKS** and **FEET**.

No opposable thumbs, no sophisticated engineering tools, not even any blueprints.

THAT'S CRAZY!

But make no mistake, some avian architects are at the top of their game.

THE
Sooty
-capped
Hermit

uses spider silk as a suspension material for her nest, and attaches the structure to a large leaf that provides rain cover. In the same way that humans design building cranes with counterweights, this ingenious bird uses a counterweight to prevent her egg-filled nest from tipping over.

THE
Village Weaver

stitches together things like GRASS, REEDS, AND PALM LEAVES to CREATE ITS WATER-WICKING Wonder nest.

THE Edible-Nest Swiflet MAKES its NESTS ENTIRELY out of its OWN SALIVA.

DROOL IS MY SUPERPOWER.

← The nest attaches to a wall!

THE MALLEEfOWL

gathers a big pile of leaves, sticks, and dirt and lays her eggs near the top. The composting leaf matter in turn warms and incubates her eggs. The male uses his beak like a thermometer, adding or removing debris to alter the temperature as needed.

BALD EAGLES

spend months building some of the largest nesting sites in the world. Nests are an average of five to six feet in diameter and four feet tall, and can weigh up to two tons.

Have you noticed that we excel at everything we do?

One of the oldest known nests in the world is on a cliff in Greenland, where

Gyrfalcons

still return to a nest that's estimated to be 2,500 years old.

Their adorable chicks balance on cliff ledges.

Cliff
Swallows
MAKE MASSIVE
COMMUNAL NESTS
ENTIRELY OUT OF MUD, HOUSING UP
to 3,000 BIRDS.

Both male and female

HORNED COOTS

make hundreds of trips to carry 1.5 tons of pebbles to their nest-building site in shallow water. When the stone pile grows to be about thirteen feet across by three feet high, the birds use aquatic vegetation to create the final nest on top of the mound.

TEAMWORK
MAKES tHE
DREAM WORK.

The smallest bird nest is made by a BEE Hummingbird.

(The Bee Hummingbird is only found in CUBA!)

At one inch across, it's just big enough to snuggle a few coffee-bean-sized eggs.

World's tiniest tiny home

SOME OF THE BIGGEST bird nests are those of the Social Weaver in SOUTHWEST AFRICA.

The birds live there throughout the year, and one known nest has been occupied for more than one hundred years. The structure can grow so heavy that it topples the very tree branches it rests on.

Not all birds build from scratch.

A White Tern

builds no nest at all,
but balances a single egg
on a BRANCH or ROCK.

I'M INTO MINIMALISM.

GILA WOODPECKERS

make their nests in holes dug out of trees or cactuses.

THE FEMALE
GReat
HØRNBILL

sequesters herself in an
already existing hole in a tree.

In a moment of inspiration, she closes off most of the
opening of the hole with mud mixed with her own poop,
leaving only a small vertical slit open. Undeterred by
these unconventional building materials,

her DEDICATED MAN WILL COME TO the POOP DOOR EVERY DAY

and deliver the food that sustains her while she incubates her eggs and raises her chicks.

HUMAN-BUILT BIRD ACCOMMODATIONS

For millennia, humans have built structures to lure birds into their lives. Dovecotes have been used all over the world as a way to house pigeons. People ate pigeon meat and eggs, and used their droppings to fertilize fields.

Assorted Dovecotes:

TINOS ISLAND, GREECE

DIEPPE, FRANCE

BANGALORE, INDIA

This dovecote is located at the summer palace of the Tipu Sultan, ruler of the Kingdom of Mysore in the 1700s. It served as housing for the Sultan's pigeon post carriers.

DOVECOTES *in* MIT GHAMR, EGYPT

In Turkey, during the time of the Ottoman Empire, people hung sophisticated birdhouses on public buildings.

For centuries,
elaborate birdhouses called

CHABUTRO

were placed in city centers of
Gujarat, India.

SZéKely GATES

are wooden gates with dovecotes carved by
members of the Hungarian population in Romania.

pigeons
go here

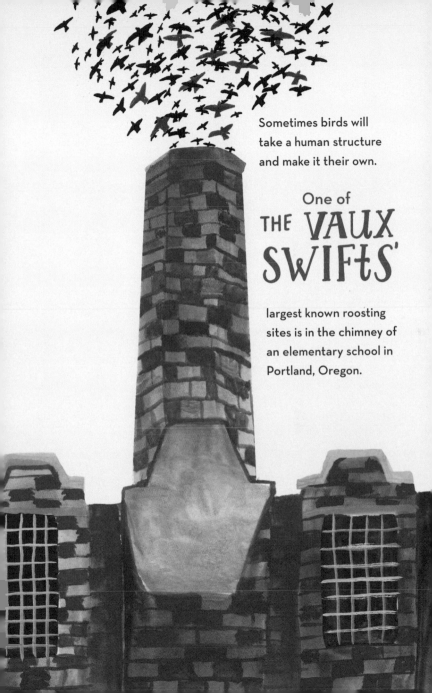

Sometimes birds will take a human structure and make it their own.

One of
THE VAUX SWIFTS'

largest known roosting sites is in the chimney of an elementary school in Portland, Oregon.

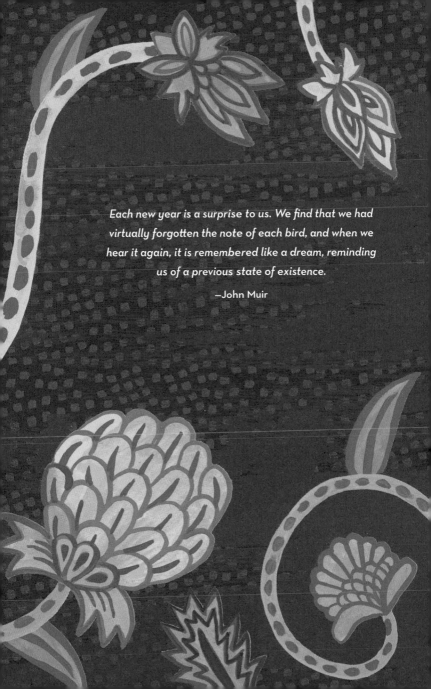

Each new year is a surprise to us. We find that we had virtually forgotten the note of each bird, and when we hear it again, it is remembered like a dream, reminding us of a previous state of existence.

—John Muir

BIRD SONG BASICS

Bird sounds are divided into two rough categories:

1. CALLS:

Basically a short single syllable, like a peep or a squawk.

HEY! YOU!

2. SONGS: A more musical and complex assortment of syllables.

"DIFFERENCE MUST BE NOT MERELY TOLERATED, BUT SEEN AS A FUND OF NECESSARY POLARITIES BETWEEN WHICH OUR CREATIVITY CAN <u>SPARK</u> LIKE A DIALECTIC." *

*AUDRE LORDE

Bird songs can be SOOTHING and BEAUTIFUL.

It's no wonder that studies show that humans feel more healthy and relaxed when surrounded by a natural soundscape that includes birdsong (even if it's a recording!).

zweeet

chuk chuk

chicker cheee

But that enchanting explosion of birdsong on a spring morning, known as the dawn chorus, is actually a testosterone-fueled free-for-all. It's made up almost entirely of male birds vying to claim dominance over a territory or to attract a mate.

BACK OFF!

Like humans, birds aren't born with "language";

THEY MUST LEARN SONGS FROM THEIR PARENTS

during the early phase of life. The specific brain circuitry that allows for vocal learning is similar in birds and humans, and new research shows that songbirds even use their own rules of grammar in their songs.

It's PRONOUNCED SEE-RAH-CHA.

SRIRACHA **HOT** CHILI SAUCE

NATURAL COLOR

SHAKE WELL

TƯƠNG ỚT SRIRACHA

是拉差香甜辣椒醬

東震食品公司

FONG FOODS, INC.

Ave. ROSEMEAD CA 91770-1114

528

WWW.HUYFONG.COM

(1 LB. 1 OZ.) (482 g) (435 mL)

FIRE-TUFTED BARBET

BIRDS USE A VOCAL ORGAN CALLED A *Syrinx* TO CREATE SOUND.

Songbirds are especially adept at controlling either half of the syrinx separately to make an incredible range of sounds. When a male uses the syrinx to create two sounds at once, it's called

A SEXY SYLLABLE.

(Yes, this is the actual scientific term.)

CRESTED
TIT

For songbirds, precise singing combined with sexiness (scientifically speaking) is a sign of health and strength. The quality of a male bird's song can literally determine if he will have success in breeding.

Birds hear more detail in their songs than we can imagine.

What may sound like a single note to our ears may actually be a burst of several quick-fire notes, or hit frequencies that we can't detect. For this reason, researchers study the timing and frequencies of birdsong visually using computer-generated images called spectrograms.

THE MALE
KAKAPO

calls lustily into the night in a way that seems strikingly unbirdlike. During breeding season, he inflates his thoracic air sac to create a very low-frequency boom that can be heard more than five kilometers (three miles) away. He makes this call all night, every night, for three months, in a single-minded attempt to attract a female *(IF HE DOESN'T ANNOY HER to DEATH FIRST).*

MAKE ROOM FOR
THE BOOM, BABY!

THE MALE
BROWN
THRASHER,

on the other hand, takes a more sophisticated approach to wooing the female.

HE HAS A REPERtOiRE of at least 1000 DISTiNCt SONGS.

YOU SHOULD HEAR ME ON SHUFFLE.

PRINCE'S TWO PET DOVES, *Majesty* & *Divinity*, WERE ALSO KNOWN FOR *their* *Enchanting songs*.

Prince credited them for "ambient singing" in the liner notes of his 2002 album *One Nite Alone.*

SADLY SOME BIRDS,
LIKE STORKS and PELICANS,
DON'T EVEN HAVE A VOICE.

I'll blink once for YES
and twice for NO.

THE MALE *superb* LYREBIRD MAKES SOME of the *most* COMPLEX SOUNDS of ANY CREATURE on EARTH.

CLICK CLICK ZOOOOM

ALSO,
HIS tAIL
FEATHERS
ARE RAD.

Besides being a perfect mimic of pretty much any other bird, he can also imitate such human-made sounds as a car engine, the click of a camera shutter, or the revving of a chainsaw.

Some birds, such as Parrots and Corvids, can mimic human speech, a skill that even our closest primate cousins (with whom we share 99 percent of our DNA!) can't accomplish.

One British Parrot that got lost from its owner returned home a few years later speaking Spanish.

SINGING with the BIRDS

IN SOME PARTS OF SOUTHERN AFRICA, HUMANS CAN COMMUNICATE WITH BIRDS.

Using clicks, whistles, and other calls,
people can summon

GREATER HONEYGUIDES,

who in turn lead them to bee colonies.

This relationship is mutually beneficial: Humans break open the hive to collect honey, and the Honeyguides enjoy the larvae-filled honeycomb. This collaboration, which has occurred for thousands of years in Tanzania, Kenya, and Mozambique, is a rare example of humans communicating with another species in the wild.

Adaptation

The radical otherness of birds is
integral to their beauty and their value.
They are always among us but never of us.

—Jonathan Franzen

Over millions of years, birds have developed specialized behaviors that allow them to survive and thrive in the wild. While nothing short of miraculous, some of these behaviors border on the macabre . . .

Otherwise *adorable*

GREAT TITS

HAVE BEEN KNOWN to SNEAK UP AND <u>MURDER</u> SLEEPING BATS BY PECKING THEIR HEADS OPEN.

I enjoy sunflower seeds and FRESH BLOOD.

THE BOREAL OWL

stocks up on a supply of rodents that he strategically leaves to freeze in cold weather. When he's hungry, he sits on one of the frozen critters until it's warm enough to eat.

NATURE'S microwave dinner!

ONE CAN ALMOST HEAR THE
Metallica soundtrack
IN THE BACKGROUND
as a
Roadrunner
BATTLES HIS FAVORITE SNACK:
RATTLESNAKE.

After his conquest, the Roadrunner will nonchalantly spend hours with part of the snake dangling out of his mouth as he waits for the rest of it to digest.

EASTERN
Screech Owls

bring small, worm-looking reptiles called blind snakes to live in their nests. The snakes burrow into the bottom of the nest and eat insects such as ants or termites, improving the health of the nest and the chance of survival for young owlets.

THE Hooded Pitohui

is one of just a few birds that have poisonous feathers and skin; they are covered in the same neurotoxin found on poisonous frogs in South America.

AND tHIS IS WHY you SHOULD NEVER LICK A BIRD.

A HOATZIN

has a digestive system that is more akin to that of cattle than to that of birds: He digests his diet of mostly leaves using a process of microbial fermentation in his stomach.

MOOO... JUST KIDDING.

He looks like a creature from an ancient legend!

Hoatzin chicks have two reptilelike claws on each wing that help them to climb up tree branches.

THE **BEARDED VULTURE** MOSTLY EATS BONES.

While it can swallow small bones whole, it will carry larger ones high into the air and drop them in order to crack them open. Its stomach is extremely acidic and can break down bones within twenty-four hours.

SALADS ARE FOR WIMPS.

Also called "QUEBRANTAHUESOS" in Spanish, which means roughly "BONE-BREAKER."

Bearded vultures are also the only birds known to intentionally decorate themselves. They use their beaks and feet to spread red mud rich in iron oxide across their bodies.

While hunting in shallow water,

THE BLACK HERON

wraps its wings around its body to form a makeshift umbrella. Fish are attracted to the shade, and the heron can see its prey better while blocking the sun's glare.

That being said, while some of these adaptations are macabre, others are simply mesmerizing.

THE
Sword-billed
Hummingbird

has the longest beak in the world relative to body size, and it's the only known bird whose beak sometimes grows longer than its body.

Incredibly, Sword-billed Hummers coevolved with a species of flower called *Passiflora mixta*. The two have a mutually beneficial relationship: The flower depends on the bird for pollination, and the bird receives a nourishing meal.

A POTOO

has small slits along the bottom of his eyelids so that he can sense movement and shadows even with his eyes closed. When a Potoo sits on a tree branch, his incredible camouflage powers make him almost impossible to see.

WHERE IS HE!?

But HE'S COMICALLY IMPOSSIBLE to MISS WHEN HIS EYES ARE OPEN.

I don't understand why you used the word "comically".

THE MARABOU STORK

DEFECATES ON ITS OWN LEGS IN ORDER TO COOL OFF.

SO REFRESHING!

THIS DAPPER NECK POUCH IS USED TO ATTRACT FEMALES.

SERVING *the* BIRDS

While birds adapt in order to survive the exigencies of their environment, we can adapt our human-designed spaces to promote bird survival as well.

SOME BIRD-FRIENDLY CHANGES YOU CAN MAKE At HOME:

- Fill your garden with native plants and trees to help restore critical bird habitat.

- Provide bird feeders and a clean water source for drinking and bathing.

- Put up birdhouses.

- Avoid chemical insecticides in your garden as much as possible.

BLACK-CAPPED CHICKADEE

- Keep cats indoors (especially during the day), or set them up with a decorative collar and bell to scare birds away.

- Cover large, clear, reflective windows in some kind of marking that deters birds from crashing into them (such as horizontal tape lines, one-way transparent film, or zen wind curtains).

- Install a chimney swift tower, or other roosting towers or platforms.

- Promote Lights Out in your city. Birds can become fatally disoriented at night by artificial building lights. Many cities, such as Boston, Chicago, San Francisco, and New York, participate in Lights Out programs. For example, in Chicago, during spring and fall migration, managers of buildings over forty stories high turn off unnecessary lighting after 11:00 p.m.

Conclusion

*We can trace our genealogy
to the origins of the universe,
and therefore rather than us being masters
of the natural world, we are part of it.*

—Gerrard Albert, Whanganui tribal leader,
New Zealand

In every moment,
BILLIONS of BIRDS ARE IN MOVEMENt
AROUND US, living out their own
GREAt <u>LIFE</u> <u>DRAMAS</u>.

Meanwhile, they are supporting our existence. Birds serve as
plant pollinators and seed distributers, and their guano makes
for the perfect fertilizer. They also help balance the ecosystem
by feeding on a variety of creatures.

But birds are in trouble. They are too often casualties of
the environmental exploitation and habitat loss that's come
with modern development. One global report says at least
40 percent of the world's bird species are in decline.

THE ATLANTIC PUFFIN:

One of many species of
birds that have been listed
as vulnerable to extinction.

But it's our very love of birds that can inspire us to change these trends. When we really connect with nature, we feel happy and fulfilled. We open ourselves to a dimension of the human experience that feels authentic and worthwhile. This connection promotes our desire to stop environmental degradation.

The first law of ecology is that everything is connected to everything else. Birds are connected to the skies, and we are connected to birds.

BIRDS CAN HELP US FIND WHO WE tRULY ARE:
MEMBERS of a DYNAMIC, INtERDEPENDENt GLOBAL FAMILY.

BIBLIOGRAPHY

Ackerman, Jennifer. *The Genius of Birds*. New York: Penguin, 2017.

Boreal Songbird Initiative. "Boreal Flyways Map," https://www.borealbirds.org/publications/boreal-flyways-map.

Chowder, Ken. "John James Audubon: Drawn from Nature." PBS, July 25, 2007, http://www.pbs.org/wnet/americanmasters/john-james-audubon-drawn-from-nature/106/.

The Cornell Lab of Ornithology, https://www.birds.cornell.edu.

Couzens, Dominic. *Extreme Birds: The World's Most Extraordinary and Bizarre Birds*. Ontario: Firefly Books, 2001.

Green, Nile. "Ostrich Eggs and Peacock Feathers: Sacred Objects as Cultural Exchange between Christianity and Islam." *Al-Masāq* 18, no. 1 (October 21, 2010): 27–78, https://www.tandfonline.com/doi/full/10.1080/09503110500222328.

Harrison, Kit, and George Harrison. *Birds Do It, Too: The Amazing Sex Life of Birds*. Minocqua, WI: Willow Creek Press, 1997.

Johnston, Alison, and Justin G. Schuetz. "Characterizing the Cultural Niches of North American Birds." *Proceedings of the National Academy of Sciences* 116, no. 22 (May 28, 2019): 10868–873, https://doi.org/10.1073/pnas.1820670116.

Jones, Ian L., and Sampath S. Seneviratne. "Mechanosensory Function for Facial Ornamentation in the Whiskered Auklet, a Crevice-dwelling Seabird." *Behavioral Ecology* 19, no. 4 (July–August 2008): 784–90, https://doi.org/10.1093/beheco/arn029.

Lawler, Andrew. *Why Did the Chicken Cross the World?: The Epic Saga of the Bird That Powers Civilization*. New York: Atria Books, 2014.

National Audubon Society, https://www.audubon.org.

Robbin, Jim. *The Wonder of Birds: What They Tell Us about Ourselves, the World, and a Better Future*. New York: Spiegel & Grau, 2017.

Sartore, Joel, and Noah Strycker. *Birds of the Photo Arc*. Washington, DC: National Geographic, 2018.

Strycker, Noah. *The Thing with Feathers: The Surprising Lives of Birds and What They Reveal about Being Human*. New York: Riverhead Books, 2014.

Stutchbury, Bridget. *The Private Lives of Birds: A Scientist Reveals the Intricacies of Avian Social Life*. London: Walker Books, 2010.

Tekiela, Stan. *Feathers: A Beautiful Look at a Bird's Most Unique Feature*. Cambridge, MN: Adventure Publications, 2014.

Young, Jon. *What the Robin Knows: How Birds Reveal the Secrets of the Natural World*. New York: Mariner Books, 2012.

PHILBY'S PARtRIDGE

PUT A BIRD ON IT.

BLUE-CROWNED
MOTMOT

BLUE
TIT

THANK *you*

I would like to thank my editor, Meg Leder, who remembered my work after many years and took a chance on this project. Also, thanks to my agent, Jennifer Weltz, who has worked with me on many different book ideas and was instrumental in making this one a reality.

I would like to acknowledge the ornithologists and zoologists whose research helps the general public understand more about birds and their habitats. One quick fun fact presented in these pages may have resulted from years of their tireless work in the field. I have a deep respect for the researchers who explore these species so that we can better understand our complex relationship with nature and create appropriate conservation strategies. If I drew a foot wrong or messed up any other specific avian detail, the mistake is mine.

Thanks to my husband, Nick, for his support. When I was trying to complete a few dozen pages close to my deadline, he suggested that I just leave them all blank and write, "Now it's your turn to draw some birds!" Deep thanks to my mom and sister, Monica, Narmin, Derik, Syda, Kelsey, Donesh, Farzad, Sabreen, Faiza, Adriana, Monica E., Gabe, Bella, Felice, Leila, Elijah, Jenny Sue, Nate, Mackenzie, Mae, and Alicia.

Finally, this one goes out to the Couch's Kingbird. A 2019 paper published in the *Proceedings of the National Academy of Sciences* ranked this cute little chirper as the least popular bird in North America, out of 621 species surveyed. The research was intended to help identify what people really value about different birds so that conservation efforts can be more strategic. So let's all give this guy a little love.

THE COUCH'S KINGBIRD:
A (very cute) BIG-TIME LOSER